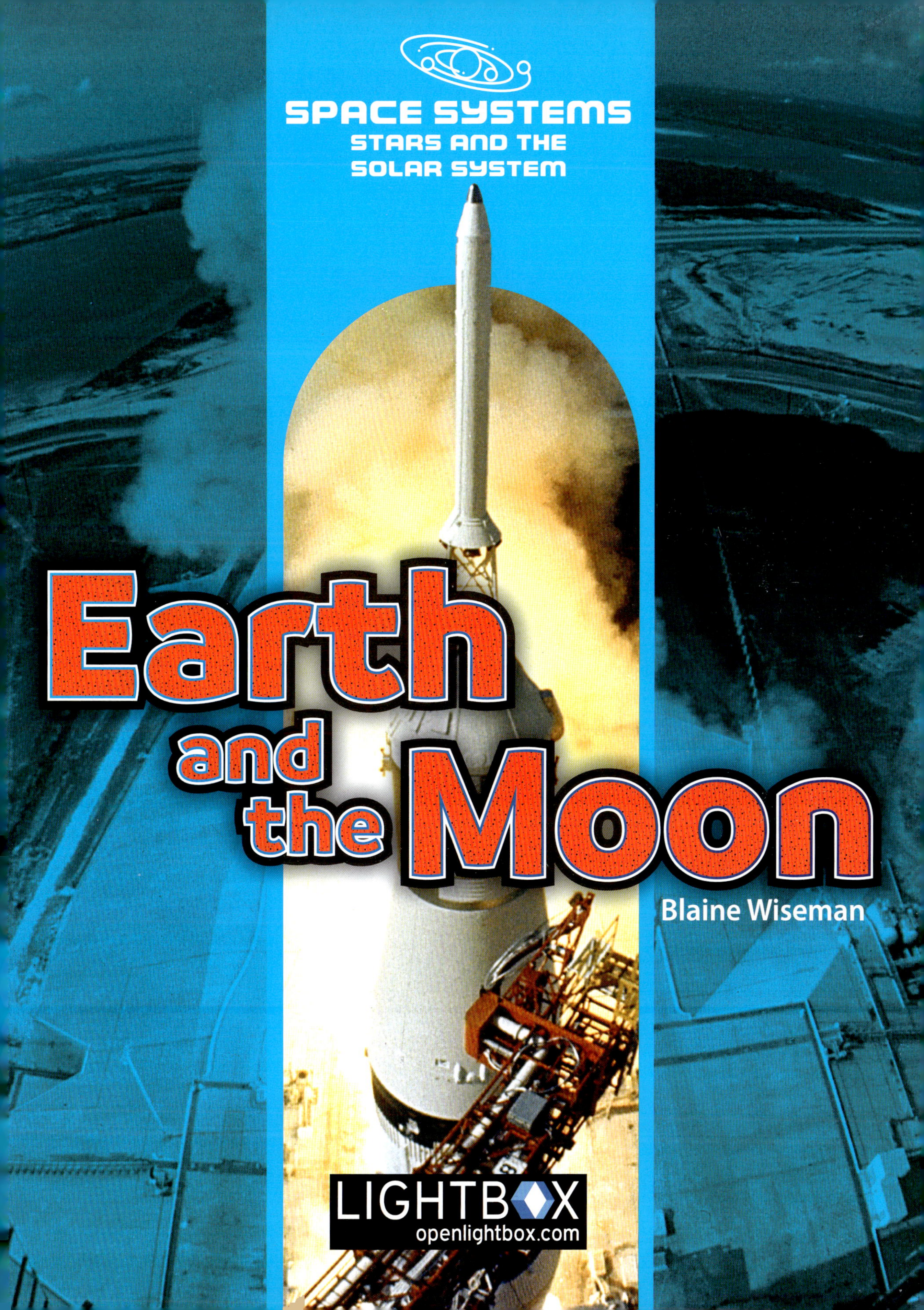
SPACE SYSTEMS
STARS AND THE
SOLAR SYSTEM
Earth and the Moon
Blaine Wiseman
LIGHTBOX
openlightbox.com

LIGHTBOX

Go to
www.openlightbox.com
and enter this book's
unique code.

ACCESS CODE

LBXW9395

Lightbox is an all-inclusive digital solution for the teaching and learning of curriculum topics in an original, groundbreaking way. Lightbox is based on National Curriculum Standards.

STANDARD FEATURES OF LIGHTBOX

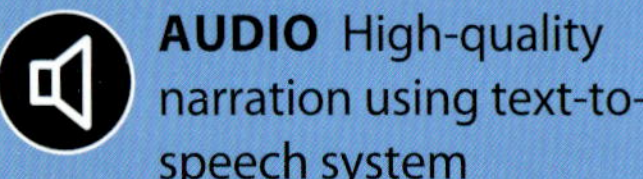

AUDIO High-quality narration using text-to-speech system

ACTIVITIES Printable PDFs that can be emailed and graded

SLIDESHOWS Pictorial overviews of key concepts

VIDEOS Embedded high-definition video clips

WEBLINKS Curated links to external, child-safe resources

TRANSPARENCIES Step-by-step layering of maps, diagrams, charts, and timelines

INTERACTIVE MAPS Interactive maps and aerial satellite imagery

QUIZZES Ten multiple choice questions that are automatically graded and emailed for teacher assessment

KEY WORDS Matching key concepts to their definitions

SPACE SYSTEMS
STARS AND THE SOLAR SYSTEM

Earth and the Moon

Contents

The Moon:
Our Night Light

Looking up into the dark night sky, humans can see thousands of stars twinkling in space. Distant planets such as Venus and Mars shine brightly enough to be seen with the naked eye. **Satellites** blink across the distance. However, the biggest, brightest night light of all is the Moon.

In reality, the Moon is not a light. It is made of rock. Light from the Moon is actually sunlight. When night falls on one part of Earth, day breaks on the other side of the world. After the Sun sets, its light cannot reach the part of Earth where it is night. Instead, sunlight shines through space and reflects off the Moon's surface. This reflection is the moonlight people see.

The Moon is Earth's only natural satellite. While the Moon seems far away, it is actually Earth's closest neighbor. The Moon is not nearly as big as most other objects people see in the sky. It only looks large because it is so close to Earth. The Moon is about 400 times smaller than the Sun, but it is also about 400 times closer to Earth.

The **Moon** formed about **4.5 billion** years ago.

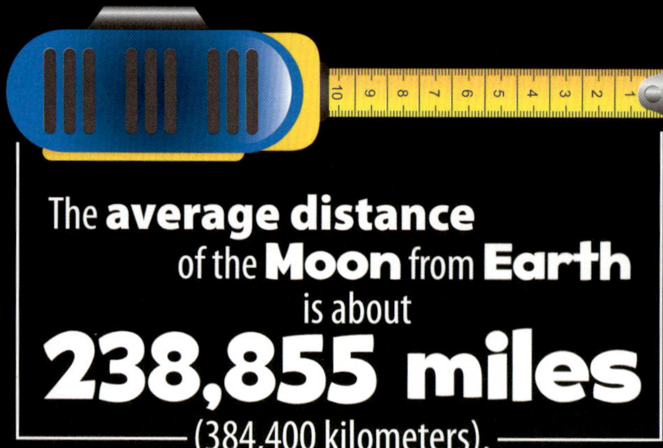

The **average distance** of the **Moon** from **Earth** is about **238,855 miles** (384,400 kilometers).

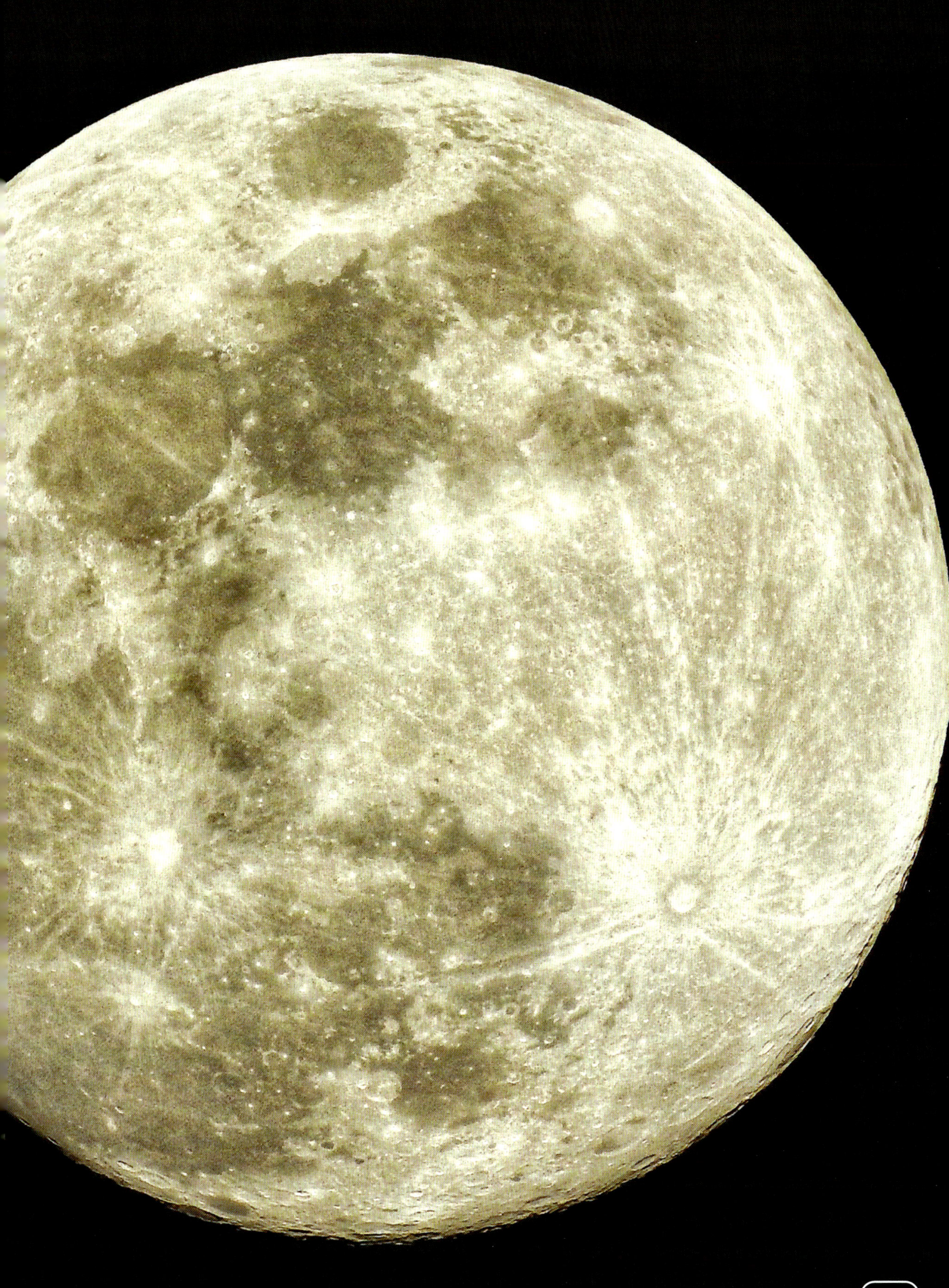

Patterns of Movement

Each night, the Moon appears to change. Its shape, brightness, and position in the sky are different from night to night. These are not actual changes to the Moon. They are changes in the way it appears from Earth. These changes are caused by the movements of both the Moon and Earth. The two major movements causing the changes are **orbit** and rotation.

The Moon orbits Earth. Earth orbits the Sun. These orbits are not perfect circles. Instead, objects orbit in an **elliptical** pattern. The Moon's elliptical orbit means that it is not always the same distance from Earth. At its **apogee**, it is about 252,088 miles (405,696 kilometers) away. During its **perigee**, the Moon may come as close as 225,623 miles (363,105 km) to Earth.

Planets and moons also rotate, or spin, while they orbit. Earth makes a full rotation every 24 hours, or each day. Earth's rotation causes the Sun, Moon, and stars to seemingly move across the sky each day and night. It makes it look like these objects move from east to west. The Moon orbits in the same direction, but at a slower rate than Earth spins. For this reason, the Moon rises in the east about 50 minutes later each night. It takes one complete lunar orbit for the Moon to begin this cycle again.

A supermoon appears when the Moon is both full and close to its perigee. It appears larger and brighter than a normal full moon.

Synchronous Rotation

One thing that does not change is the side of the Moon that is visible from Earth. It takes about 27 days for the Moon to make one full rotation. It also takes the Moon about 27 days to orbit Earth. This is called **synchronous** rotation. For this reason, the same side of the Moon is always facing Earth.

Lunar Timeline

Since ancient times, humans have looked up in wonder at the Moon. Throughout history, new ways of observing the Moon have been developed. Today, humans even have the technology to travel to the Moon. Scientists and astronauts are now studying ways to use the Moon as a base to reach deeper into space.

1609 AD

Italian **astronomer** Galileo Galilei becomes the first person to use a telescope to observe the Moon. He discovers that the Moon's surface is not smooth, but covered in craters and mountains.

About 3000 BC

Ancient people in Ireland carve the world's oldest known map of the Moon. The stone carving shows the Moon's visible features. On certain nights, the light of a full moon shines down a passage onto the carving.

2019

A new space program, known as Artemis, is created by NASA. By 2024, Artemis III will land the first woman astronaut on the Moon. In Greek mythology, Artemis is the goddess of the Moon and the twin sister of Apollo.

1961

In a speech to the U.S. Congress, President John F. Kennedy says, "This nation should commit itself to achieving the goal, before this decade is out, of landing a man on the Moon and returning him safely to Earth."

1969

The National **Aeronautical** and Space Administration (NASA) achieves Kennedy's goal. On the Apollo 11 mission, astronauts Neil Armstrong and Buzz Aldrin become the first people to walk on the Moon. Ten more astronauts would walk on the Moon during later Apollo missions.

1958

The United States launches Pioneer 0, the first spacecraft designed to orbit the Moon. It is an **unmanned** mission. The mission fails when Pioneer 0 explodes 73 seconds into its flight.

Gravity and the Moon

Orbiting is caused by gravity. All objects with mass have gravity. This includes Earth, the Moon, the Sun, a building, a brick, a feather, and a person. Gravity pulls things toward an object's center of mass. It is the same force that makes things fall down instead of up. A larger object has a stronger gravitational pull than a smaller one. The solar system, including Earth and the Moon, is pulled in orbit by the Sun. The Moon and Earth also pull each other.

Gravity on Earth is about six times stronger than it is on the Moon. This means that, on the Moon, an object weighs about one sixth of what it weighs on Earth. Astronauts on the Moon have noticed the effects of weaker gravity. For example, they are able to jump much farther there than on Earth.

Gravity and the Tide

Tides are the rise and fall of water levels in Earth's oceans and other large bodies of water. As the Moon passes around Earth each day, its gravity pulls water in the Moon's direction. This causes the world's oceans to bulge. Waters closest to the Moon experience **high tide**. The bulge caused by gravitational pull also creates high tide on the opposite side of the world. This causes a **low tide** in areas that are not part of the bulge.

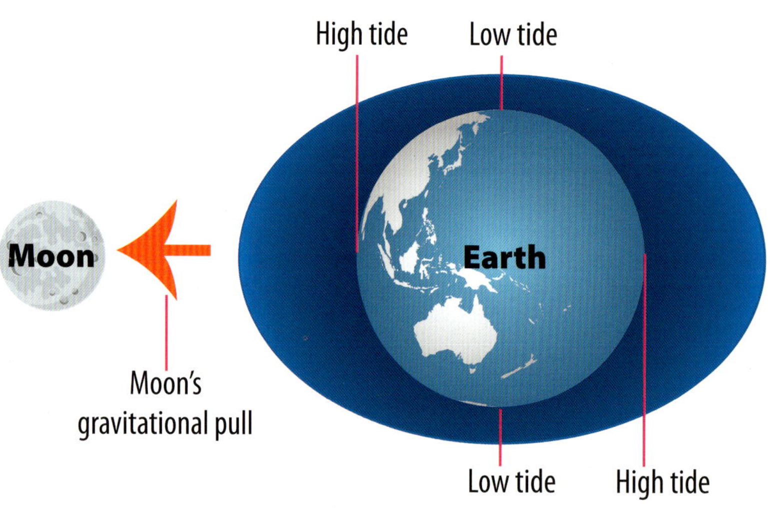

Sun

Astronaut Alan Shepard brought a golf club and golf ball to the Moon. He said that his shot traveled for "miles and miles and miles."

Observing the Moon

Unlike the Sun, the stars, and the planets, the Moon can easily be observed with the naked eye. People can even see features on the Moon's surface, including highlands and areas known as **seas**. These are what cause the light and dark patches on the Moon. Many people believe these features resemble a face. This is known as the "Man in the Moon." People use technology such as telescopes, zoom lenses, and even binoculars to take a closer look at the near side of the Moon.

Observing the far side of the Moon is more complicated. In 1959, the **Soviet Union**'s Luna 3 became the first spacecraft to photograph this area. Thanks to space exploration, there are now detailed maps of the Moon's entire surface. In 2019, China's Chang'e 4 lunar **lander** became the first spacecraft to **soft-land** on the far side of the Moon. It carried the Yutu-2 **rover**, which explored the Moon's surface.

Map the Moon

Create your own map of the Moon by following these steps:

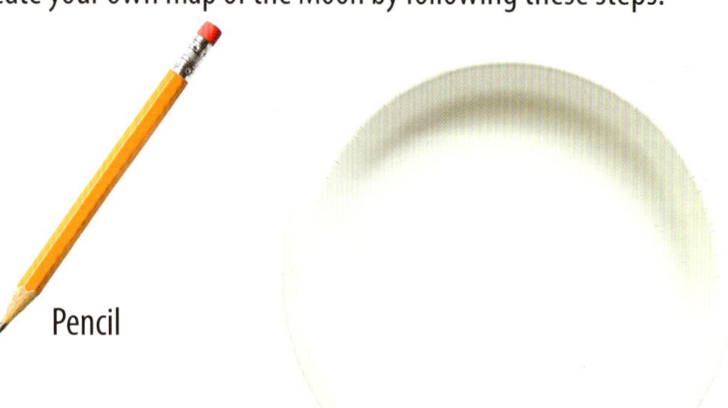

Pencil

Paper plate

Full moon

1. Look up at the night sky during a full moon. On a paper plate, draw a map of the Moon's surface using a pencil. Use shading to make the dark parts of the Moon darker on the plate.

Spacecraft, including NASA's Lunar Reconnaissance Orbiter (LRO), orbit the Moon, taking pictures and measurements. The LRO has helped scientists create highly detailed maps of the Moon's entire surface.

2. When you have finished drawing your map, use the image to the right to label the areas you drew.

Label as many of the following areas on your map as you can: Sea of Tranquility, Sea of Serenity, Sea of Showers, Ocean of Storms, Sea of Moisture, Sea of Clouds, Sea of Vapors, Sea of Nectar, Sea of Fecundity, Sea of Crises, Sea of Cold.

Mapping Lunar Launch Sites

The United States, the Soviet Union, and China have all successfully soft-landed spacecraft on the Moon. Other countries or regions, such as Japan, India, Israel, and the **European Union**, have **hard-landed** or crashed spacecraft into the Moon's surface. Only American astronauts have walked on the Moon.

Arctic Ocean

North America

Atlantic Ocean

EQUATOR

Pacific Ocean

South America

Southern Ocean

Kennedy Space Center (KSC), United States

KSC was built in Florida beginning in 1962. It became NASA's main launch site for space missions. All 12 astronauts to walk on the Moon lifted off from KSC. It was also the launch site for each of NASA's space shuttle missions. KSC was named after President Kennedy in 1963, one week after his death.

Legend
Land
Water
N W E S
Scale 0 2,000 Miles 2,000 Kilometers

Baikonur Cosmodrome, Kazakhstan

Baikonur Cosmodrome has been the launch site for all Soviet and Russian space missions throughout history. Both the first satellite and the first person to reach space lifted off from Baikonur. Baikonur is the launch site for Soyuz rockets that fly astronauts and supplies to the International Space Station.

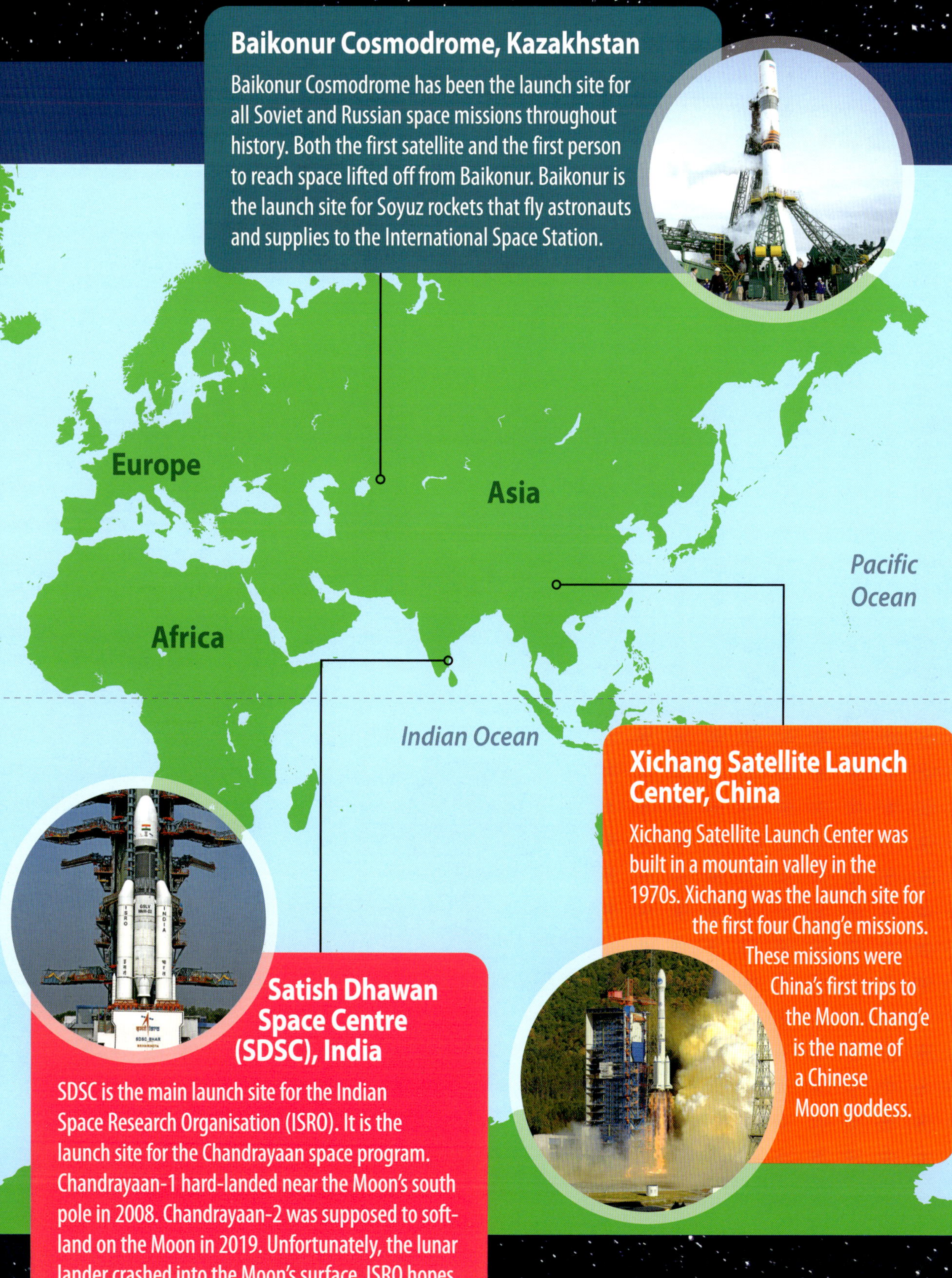

Xichang Satellite Launch Center, China

Xichang Satellite Launch Center was built in a mountain valley in the 1970s. Xichang was the launch site for the first four Chang'e missions. These missions were China's first trips to the Moon. Chang'e is the name of a Chinese Moon goddess.

Satish Dhawan Space Centre (SDSC), India

SDSC is the main launch site for the Indian Space Research Organisation (ISRO). It is the launch site for the Chandrayaan space program. Chandrayaan-1 hard-landed near the Moon's south pole in 2008. Chandrayaan-2 was supposed to soft-land on the Moon in 2019. Unfortunately, the lunar lander crashed into the Moon's surface. ISRO hopes to launch Chandrayaan-3 from SDSC in 2021.

Lunar Eclipses

Lunar eclipses happen when a shadow is cast over the Moon. Earth travels between the Sun and Moon and blocks the Sun's light from reaching the Moon. A lunar eclipse can only happen during a full moon. There are three types of lunar eclipses. The type of eclipse depends on which part of Earth's shadow falls on the Moon.

A shadow has two parts. The umbra is the dark, middle part. The penumbra is the lighter outer edge. A penumbral eclipse is hard to see. It happens when just the penumbra of Earth's shadow falls over the Moon. A partial eclipse is when part of the umbra shades the Moon. It looks like a bite has been taken out of the Moon. Total eclipses are the rarest type. The entire moon is cast in Earth's shadow. A total lunar eclipse happens over several hours as the Moon, Earth, and Sun move into a perfectly straight line. When the Moon is completely covered by shadow, the eclipse has reached **totality**.

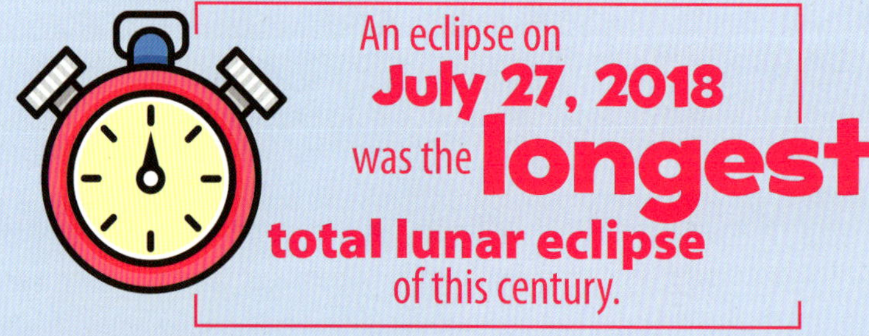

During an eclipse, the Moon does not go completely dark. Some light still reaches its surface. This light is refracted through Earth's atmosphere. The refracted light can appear red, which makes the Moon appear to be a reddish color. For this reason, some lunar eclipses are known as blood moons.

Phases of the Moon

As the Moon and Earth move through their orbits, the Moon passes through eight phases. While the Moon itself does not change, its position in space does. So does the way people see it from Earth. It takes about 29.5 days for the Moon to complete a cycle of its phases. This lunar cycle is the basis for the time period called a month.

Sunlight only reaches one half of the Moon at a time. During a full moon, all of that light is visible from Earth because it is shining on the near side of the Moon. During a new moon, the Sun's light shines on the far side of the Moon, which cannot be seen from Earth. The eight phases of the Moon are defined by the amount of light visible from Earth and the direction in which light is moving.

What begins as a small sliver of light grows from one night to the next. At first, the light appears as a crescent shape. This is caused by the Moon's round shape and the way it blocks the light. The way the light spreads across the face of the Moon is called **waxing**. The Moon continues waxing until it becomes **gibbous**, and then full. Once the Moon is full, the process of **waning** begins. Light wanes from the visible surface of the Moon until the next new moon.

Gibbous can mean "rounded" or "swelling." Gibbous moons take place between full moons and quarter moons.

Lunar Phases

Each of the eight phases of the Moon has a unique name and appearance that people use to identify it.

New Moon
The Moon appears dark, as none of the light it reflects is visible from Earth.

Waxing Crescent Moon
Light begins to spread across the Moon's visible surface.

Waning Crescent Moon
The Moon is between last quarter and new.

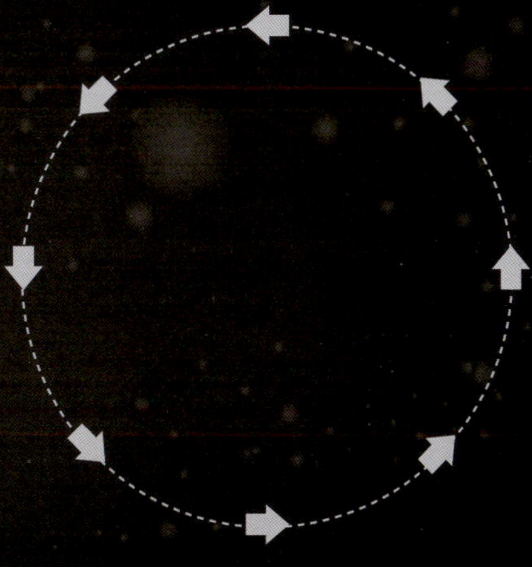

First Quarter Moon
The Moon is one quarter of its way through orbit. This phase is also called half moon because half of the Moon's visible surface is illuminated.

Last Quarter Moon
The Moon appears similar to a first quarter moon, except its other half is lit up.

Waxing Gibbous Moon
A waxing gibbous moon is between first quarter and full.

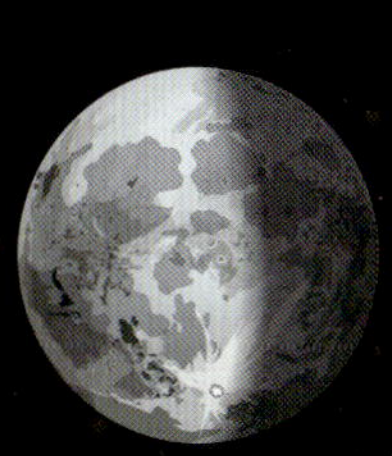

Waning Gibbous Moon
The Moon starts to become covered in shadow.

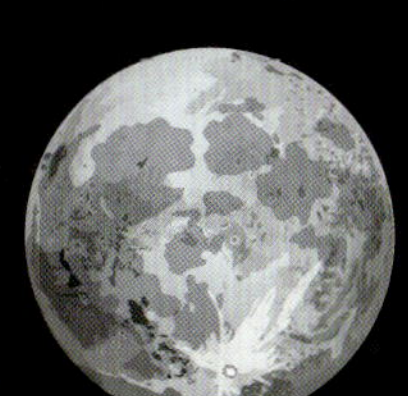

Full Moon
The Moon is halfway through its cycle. This positioning is called opposition. From Earth, the Moon appears as a perfect, fully lit circle.

Lunar Calendar

You can track the Moon's progress through its phases by making a lunar calendar. Begin recording on any day. Continue recording for the next 30 days to see one complete lunar cycle.

Materials:
Graph paper
Pen or pencil

1. Using your pen or pencil, make a monthly calendar. Divide the paper into 30 squares.
2. Each night, look at the Moon.
3. In that night's square, draw a picture of the Moon's shape.
4. Make notes of things you observe each night. For example, which phase is the Moon in? Have you seen an eclipse? How bright or dim is the Moon's light? What time are you observing the Moon? Where is the Moon in the sky?

sun mon tue wed thu fri sat
1 2
3 4 5 6 7 8 9
10 11 12 13 14 15 16
17 18 19 20 21 22 23
24 25 26 27 28 29 30

Quiz

1 What source of light causes the Moon to light up?

A: The Sun

2 Who was the first person to use a telescope to look at the Moon?

A: Galileo Galilei

3 Which spacecraft took the first pictures of the far side of the Moon?

A: Luna 3

4 What is the name of NASA's main launch site?

A: Kennedy Space Center (KSC)

5 What two parts of a shadow cause eclipses?

A: Umbra and penumbra

6 What word describes the point when the Moon is closest to Earth?

A: Perigee

7 Which astronaut played golf on the Moon?

A: Alan Shepard

8 What is the name for the lunar phase when the Moon is between first quarter and full?

A: Waxing gibbous

Key Words

aeronautical: the study of flight

apogee: most distant point

astronomer: a scientist who studies space

elliptical: an oval shape

European Union: a political group of countries in Europe

gibbous: bulging at both edges

hard-landed: landed in an uncontrolled manner, or crash landed

high tide: when water is at its highest along a shoreline

lander: a spacecraft designed to land on a surface

low tide: when water is at its lowest along a shoreline

perigee: nearest point

orbit: the path one object in space takes around another

rover: a spacecraft designed to land on and explore objects in space

satellites: things that orbit something else

seas: names given to vast areas of low land on the Moon

soft-land: land in a controlled manner

Soviet Union: a political group of countries in Eastern Europe that split apart in 1991

synchronous: moving at the same rate

totality: something that is whole or entire

unmanned: a spacecraft with no people on board

waning: decreasing

waxing: increasing

Index

SUPPLEMENTARY RESOURCES

Click on the plus icon found in the bottom left corner of each spread to open additional teacher resources.

- Download and print the book's quizzes and activities
- Access curriculum correlations
- Explore additional web applications that enhance the Lightbox experience

LIGHTBOX DIGITAL TITLES

Packed full of integrated media

VIDEOS

INTERACTIVE MAPS

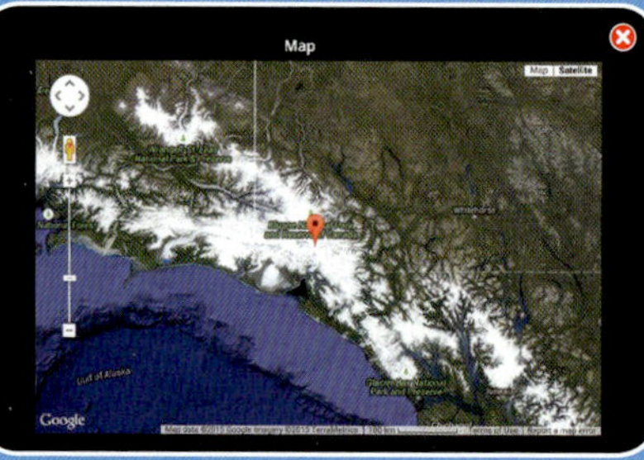

WEBLINKS

SLIDESHOWS

QUIZZES

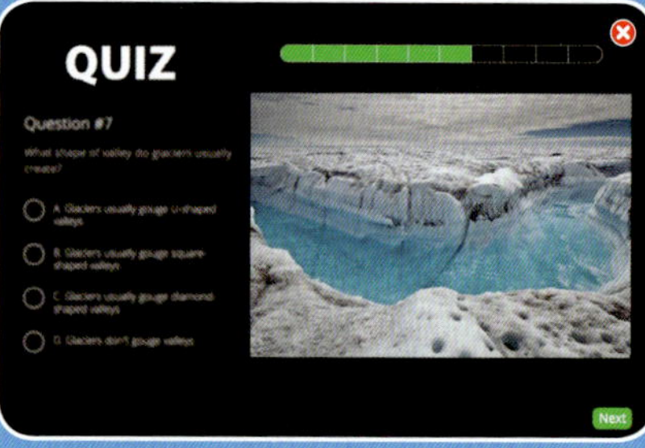

OPTIMIZED FOR

- ✓ TABLETS
- ✓ WHITEBOARDS
- ✓ COMPUTERS
- ✓ AND MUCH MORE!

Published by Smartbook Media Inc.
14 Penn Plaza, 9th Floor New York, NY 10122
Website: www.openlightbox.com

Library of Congress Cataloging-in-Publication Data

Names: Wiseman, Blaine, author.
Title: Earth and the moon / Blaine Wiseman.
Description: New York, NY : Lightbox/Smartbook Media, [2021] | Series: Space systems: stars and the solar system | Includes index. | Audience: Ages 9-14 | Audience: Grades 4-6
Identifiers: LCCN 2020018631 (print) | LCCN 2020018632 (ebook) | ISBN 9781510554689 (library binding) | ISBN 9781510554696
Subjects: LCSH: Moon--Juvenile literature. | Earth (Planet)--Juvenile literature.
Classification: LCC QB582 .W568 2021 (print) | LCC QB582 (ebook) | DDC 523.3--dc23
LC record available at https://lccn.loc.gov/2020018631
LC ebook record available at https://lccn.loc.gov/2020018632

Printed in Guangzhou, China
1 2 3 4 5 6 7 8 9 0 24 23 22 21 20

062020
111019

Project Coordinator John Willis **Designer** Ana María Vidal

Photo Credits
Every reasonable effort has been made to trace ownership and to obtain permission to reprint copyright material. The publisher would be pleased to have any errors or omissions brought to its attention so that they may be corrected in subsequent printings. The publisher acknowledges Getty Images, iStock, Newscom, Shutterstock, and Wikimedia as its primary image suppliers for this title.